WHAT DOES IT DO?
Inventions Then and Now

Picture Credits

The publisher believes that every effort has been made to secure permission to reprint photographs and illustrations for this book. In the event of any omission, the publisher expresses regret and will be pleased to make the necessary corrections in future editions.

Front Cover: George Haling/Photo Researchers

Page 4 (t) Courtesy of The Minolta Corporation; page 4 (b) Suzanne Szasz/Photo Researchers; pages 4-5 Wayne Lukas—National Audubon Society/Photo Researchers; page 5 (t) Kent & Donna Dannen/Photo Researchers; page 5 (b) Charles Gupton/The Stock Market; page 7 The Bettman Archive; page 8 (t) David Frazier/The Stock Market; page 8 (b) Janeart, Ltd./The Image Bank; pages 8-9 F.B. Grunzweig/Photo Researchers; page 9 (r) Jeffrey W. Myers/The Stock Market; page 9 (b) Sobel-Klonsky/The Image Bank; page 10 The Bettman Archive; page 11 (t) Picture Collection, The Branch Libraries, The New York Public Library; page 11 (b) The Bettman Archive; page 12 (t) Courtesy of the Raleigh Cycle Company of America; page 12 (b) Lawrence Migdale/Photo Researchers; pages 12-13 Jerry Wachter/Focus on Sports; page 13 (r) J. Barry O'Rourke/The Stock Market; page 16 (t) John R. Maher/The Stock Market; page 16 (b) Adelheid Heine-Stillmark/The Image Bank; pages 16-17 Walter Iooss/The Image Bank; page 17 Franco Fontanay/The Image Bank; page 19 The Bettman Archive; page 20 (t, l) Courtesy of Chevrolet; page 20 (t, r) Philip Mackey/The Stock Market; page 20 (b) Michael Melford/The Image Bank; page 21 (t, l) Brett Froomer/The Image Bank; page 21 (t, r) Jake Rajs/The Image Bank; page 21 (b) Marc Romanelli/The Image Bank; page 22 Picture Collection, The Branch Libraries, The New York Public Library.

First Steck-Vaughn Edition 1992

Library of Congress number: 90-8014

Library of Congress Cataloging in Publication Data.

Jacobs, Daniel (Daniel Martin).
  What does it do?: inventions then and now / by Daniel Jacobs.

  (Ready-set-read)
  Summary: Compares the appearance and use of common machines such as camera, clock, telephone, bicycle, auto, train, and plane with that of the models used long ago.
  1. Inventions—Juvenile literature. [1. Inventions.] I. Title. II. Series.
  T48.J23   1990   609—dc20   90-8014

ISBN 0-8172-3586-8   hardcover library binding

ISBN 0-8114-6748-1   softcover binding

  3 4 5 6 7 8 9   96 95 94 93 92

WHAT DOES IT DO?

Inventions Then and Now

by Daniel Jacobs

RSVP

RAINTREE
STECK-VAUGHN
P U B L I S H E R S
The Steck-Vaughn Company

Austin, Texas

This is a camera.

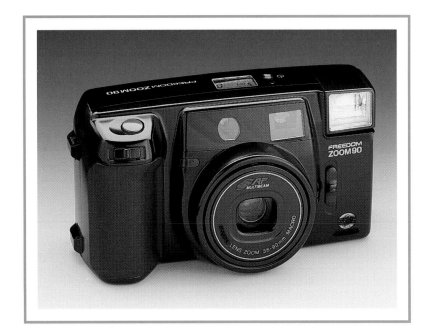

What does it do?

4

Click, click.
It takes pictures.

5

Long ago, cameras looked like this.

To have their pictures taken, people had to stand
still for five minutes. Five minutes is a long time
to stand still!

This is a telephone.

What does it do?

Ring, ring.
It lets you talk to
someone far away.

9

Long ago, telephones looked like this.

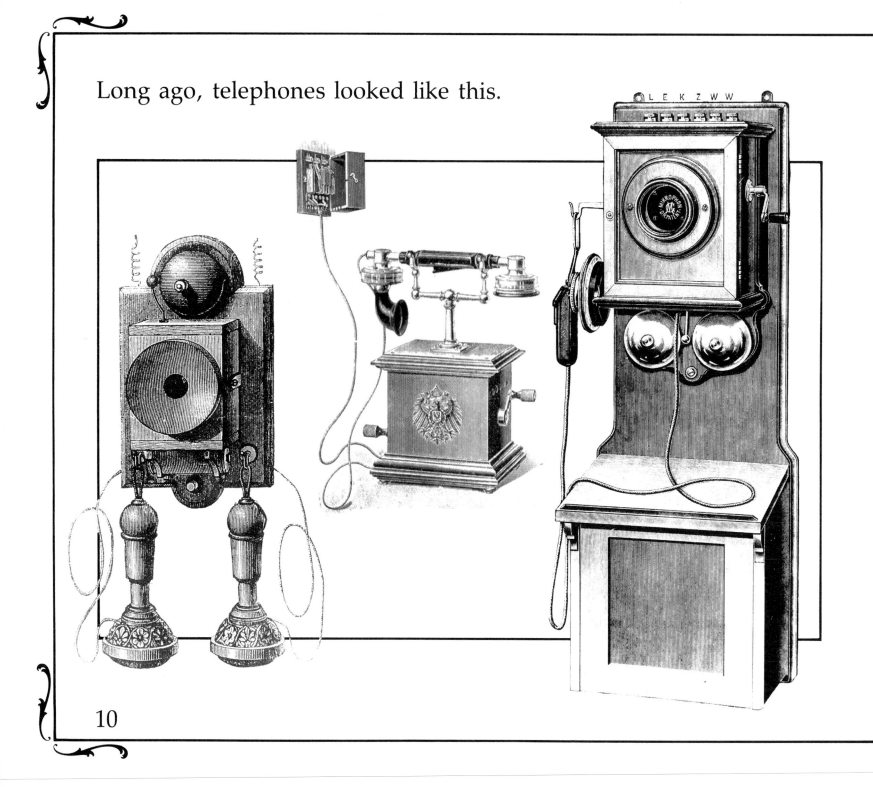

On this early telephone, people could not call someone who was very far away.

This is a bicycle.

What does it do?

Zip, zip.
It helps you go fast.

Long ago, bicycles looked like this.

This bicycle was called a boneshaker.
It gave a very bumpy ride.

This is a clock.

What does it do?

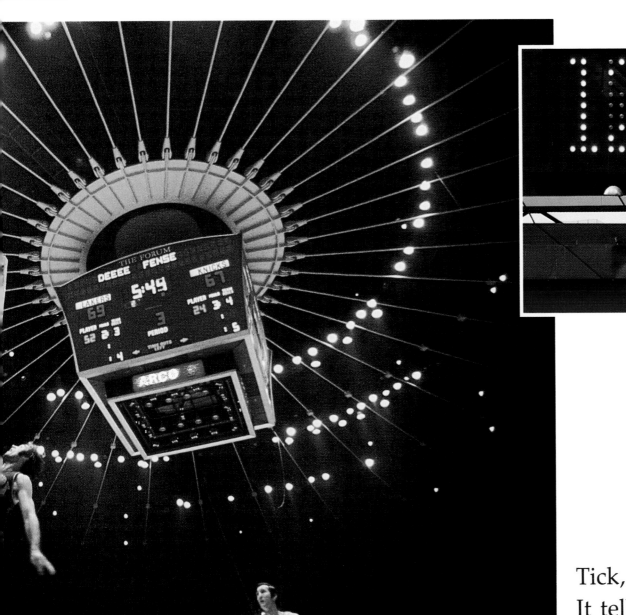

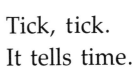

Tick, tick.
It tells time.

Long ago, clocks looked like this.

Sundials told time by measuring the shadow cast by the sun. This made it hard to tell time on a cloudy day.

This is a car.  This is a plane.

This is a train.

What do they do?

Zoom, zoom.
They take you from
place to place.

21

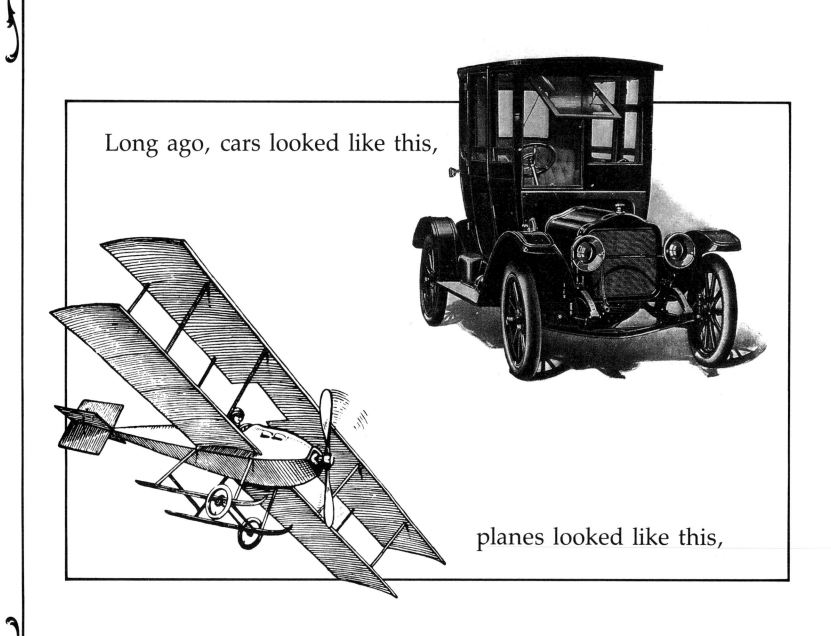

Long ago, cars looked like this,

planes looked like this,

and trains looked like this.

Things sure were different long ago!

Sharing the Joy of Reading

Reading a book aloud to your child is just one way you can help your child experience the joy of reading. Now that you and your child have shared **What Does It Do?**, you can help your child begin to think and react as a reader by encouraging him or her to:

- Retell or reread the story with you, looking and listening for the repetition of specific letters, sounds, words, or phrases.

- Make a picture of a favorite character, event, or key concept from this book.

- Talk about his or her own ideas and feelings about the subject of this book and other things he or she might want to know about this subject.

Here is an activity that you can do together to help extend your child's appreciation of this book: The telephone is one of the machines described in **What Does It Do?** You can help your child learn how to make phone calls and how to answer the phone correctly. Together, you might want to make a list of important and emergency phone numbers to keep near the phone.